Gene Expression
And Its Regulation

Gene Expression And Its Regulation

LAYING THE FOUNDATION
FOR MOLECULAR BIOLOGY
A PERSONAL ACCOUNT

Werner Maas

Rev. date: 02/27/2013

To order additional copies of this book, contact:
Xlibris Corporation
1-888-795-4274
www.Xlibris.com
Orders@Xlibris.com
128175

Table of Contents

Chapter 1

How I Decided to Study Biochemical Genetics

In September 1939, I arrived in Cambridge to join the class of 1943 of Harvard College. Previously, I had decided to major in biology. This had come about by a process of elimination of other possibilities. I would have liked to become a musician or a writer, but I felt that I did not have enough talent to make a living in these fields. Business or law was out of the question. I decided against medicine because I had seen that my father, who was a physician, in his active practice, had never had time for medical studies. Chemistry and physics were out because they required too much mathematics. This left me with biology, which I liked, as a possible career for my future.

During the first two years at Harvard, I found biology quite dull, except for a course in comparative anatomy given by Alfred Romer, an inspiring teacher. It was only in the third year that I became excited about biology, and this came about as follows. During the first semester, I took a course in genetics, which opened a totally new aspect of biology to me, of whose existence I had not been aware. Previous biology courses had been mostly descriptive without much underlying theory, except for the theory of evolution. In contrast, here I encountered abstract units of inheritance called genes. They could be treated quantitatively, and their transmission from one generation to the next was governed by rules that permitted one to predict the frequencies of different types of offspring. To me, biology became a real science.

However, there was something about the genetics course that bothered me. Why was nothing said about the chemical constitution of genes and the reactions they carried out leading to their products? This question became more urgent during the next semester when I took a course in biochemistry. The teacher, George Wald, a brilliant lecturer, gave lucid descriptions of metabolic pathways but never talked about genes. I came under the impression that biochemists had never heard of genes and geneticists had never heard of enzymatically controlled reactions. At this point, I decided to investigate the biochemical basis of gene action. This decision was the starting point for my future research. Unknown to me, the field of biochemical genetics had been launched shortly before this time.

In 1941, the geneticist George Beadle and the biochemist Edward Tatum published their first studies on mutants of the bread mold *Neurospora* in which the biochemical action of the mutated reaction was known. This epoch-making paper was entitled "Genetic Control of Biochemical Reactions in *Neurospora*." On the basis of the results of this study, Beadle proposed subsequently his famous one gene—one enzyme hypothesis.

Prior to the studies with *Neurospora*, Beadle and Tatum had investigated the problem of gene action with mutants affecting one of the eye pigments of the fruit fly *Drosophila* and had shown that each step in the formation of this eye pigment is controlled by a single gene. They had found this system to be cumbersome for studies of gene action and had switched to the much more suitable bread mold.

In 1942, I looked for a laboratory where I could carry out experiments for my senior honor thesis related to gene action. I was delighted to learn that George Wald had a student who was working on the chemistry of *Drosophila* eye pigments. I asked Wald if I could join his student, and he accepted me. The student, Gordon Allen, was a year ahead of me, and I continued this work after he graduated (with summa cum laude!). Gordon later became a human geneticist at the National Institutes of Health, and we have remained lifelong friends.

During my work on eye pigments, I learned two basic techniques that were useful in my later research: spectrophotometry, to measure light absorbed by colored solutions, and chromatography, to separate component pigments by passing the mixture through a column of an adsorbent powder. Wald was very good in explaining these procedures and in relating the history of their discovery and development.

At the end of my stay in his laboratory, Wald invited me to a New Year's Eve party at his home. "There is a biochemist friend I first met during my postdoctoral stay in Berlin I want you to meet. His name is Fritz Lipmann, and he is a genius."

The party was lively and animated except for one person who remained silent for the whole evening, deeply absorbed in his thoughts. This was Fritz Lipmann. At the time, he was studying how high-energy phosphate bonds he used to drive the biosynthesis of cellular components.

This was my first encounter with Lipmann, who, as we shall see, later became a very important person in my life.

I graduated from Harvard College in 1943 after the first semester. Subsequently, I obtained a fellowship and started graduate school at Harvard. During the second semester, I took bacteriology, which was given to second-year medical students. For this course, I had to go to the medical school in Brookline. I went there on my bicycle. The subject was interesting, but I felt out of place, being the only graduate student among the medical students. The laboratory instructor was aware of my situation and paid special attention to me. His name was John Enders. He later won a Nobel Prize for his fundamental work toward developing a polio vaccine.

At the end of the semester, I decided to leave Harvard and go to a university that was more oriented toward genetics. I would have liked to move to Stanford and join Beadle's group, but it was impractical for me to move to California. I decided to apply to Columbia University in New York City. It was a stronghold of genetics, with the eminent geneticists L. C. Dunn, Theodosius Dobzhansky, and M. Rhoades on its faculty. I was interviewed by Professor Dunn, whose textbook we had used in the genetics course at Harvard. After that, I was accepted as a graduate student in the Department of Zoology and offered a position as a teaching assistant. I was very happy to go to Columbia, especially since I could live at home in my mother's apartment.

Chapter 2

Five Student Years of Preparation

Figure 1. Theodosius Dobzhansky.

Figure 2. The author, left, talking to Professor T.M. Sonneborn
at the 1947 Cold Spring Harbor Symposium.

In 1943, there were circumstances that made me want to finish my graduate studies and obtain a PhD degree as quickly as possible. World War II was in progress, the graduate student body was depleted, and there was a general feeling of urgency to complete one's education.

I had two problems to solve that were related to each other. One was to fulfill the sixty points of required course work and finish a research project for the PhD in a reasonably short time. The other was to find a faculty sponsor for my PhD project whose research was within the field of biochemical genetics. In the absence of such a person, I decided that I could solve my problems by continuing my research on *Drosophila* eye pigments with Dobzhansky as a sponsor since he worked with *Drosophila*. For the chemical part of my research, I would look for a sponsor in the Chemistry Department. My plan was accepted by the faculty. In the Chemistry Department, Professor Charles Dawson, an organic chemist who was working on aromatic compounds with biological activity, such as the active principle of poison ivy, offered me space in his laboratory and agreed to advise me in my experiments. With these arrangements, I managed to obtain a PhD degree in three years. The title of my thesis was "Spectrophotometric and Chromatographic Adsorption Analysis of the Red Eye Pigment of *Drosophila melanogaster*." For my course work, I concentrated on genetics but also included some chemistry.

During the first year, I took Dobzhansky's course on fundamentals of genetics. He was a dynamic and forceful lecturer. He had the ability to explain complex subject matter in simple and concise terms. At the end of this course, I felt that now I really knew genetics.

During the same year, I took Selig Hecht's course on general physiology. Here was dynamism of a different kind but equally forceful. He dealt mainly with basic biochemical mechanisms. His lectures were spellbinding and dramatic, and I could see where his student Wald had learned his style of lecturing.

During the third year, I took Marcus Rhoades's course in cytogenetics. He was located in the Botany Department. He had been a graduate student of R. A. Emerson at Cornell. Together with two other graduate students, Barbara McClintock and George Beadle, he had been involved in working out the cytogenetics of maize. This was after *Drosophila*, the most important organism for genetic studies, and Emerson's laboratory was the counterpart of Morgan's laboratory in elucidating the chromosomal basis of inheritance. Rhoades's lectures were full of precise information, but the best part of his course was the beautiful microscope slide demonstrations of maize chromosomes.

There was a required course in general zoology. It was taught by Francis Ryan, the youngest member of the faculty who had received his PhD in 1941 in the Zoology Department. He was well grounded in the field and was a competent teacher. He also taught the basic biology course for undergraduates, and I was the laboratory instructor in this course. Besides being a zoologist, Francis had developed an interest in biochemical genetics and had gone for a year to Beadle's laboratory at Stanford.

I also took a course in physical chemistry during my first year. I had avoided taking this course at Harvard because I thought my mathematical preparation was inadequate. Actually, I got along well in this course. My teacher, Jerry Rosenberg, was a young member of the Chemistry Department and explained the subject matter very clearly. Later I found out that at this time, he was also Professor Urey's assistant in his work on the Manhattan Project.

Besides the basic courses, I took three advanced courses and a seminar course in genetics, all given by Dobzhansky. In the seminar course, the students presented reviews of important papers. We reviewed two papers that became classics in the new field of molecular biology. One was the 1943 paper by S. E. Luria and M. Delbrück on

"Mutations in Bacteria from Virus Sensitivity to Virus Resistance," which demonstrated that bacteria have genes like higher organisms and are therefore suitable for genetic experiments. Such mutants turned out later to be very advantageous for studies in biochemical genetics. For this work, Luria and Delbrück were awarded the Nobel Prize in 1958. The other paper published in 1944 was by D. T. Avery, C. M. Macleod, and M. McCarty on "Studies on the Chemical Nature of the Substance Inducing Transformation of Pneumococcal Types," and it showed that genes were made of deoxyribonucleic acid (DNA). Dobzhansky, although a population geneticist, had realized the significance of these papers when they appeared, and he assigned them to students for oral presentations, which would be thoroughly discussed in class. Thanks to him, I became familiar with these advances on the ground floor.

As Dobzhansky was a central figure for me at Columbia, I shall briefly digress to describe his education and prior career as background and explain my relationship with him (fig. 1).

Dobzhansky, born in 1900, grew up in Kiev in the Ukraine. As a youth, he was an ardent naturalist and a collector of butterflies, beetles, and ladybugs. He graduated from the university in Kiev in 1921 and stayed there as a lecturer in biology until 1924. During this time, he also did practical work, studying diseases of the sugar beet plant. He then moved to Leningrad where he was lecturer in genetics from 1924 to 1927. In 1926, he led an expedition to Central Asia to study domesticated animals.

In the early 1920s, he learned with enthusiasm about the work of Morgan's group at Columbia University, and when he was in Leningrad, he started to do experiments with *Drosophila* under the guidance of Iuril Filipchenko, an eminent zoologist. In 1927, with Filipchenko's support, he received a fellowship from the Rockefeller Foundation to work in Morgan's laboratory for two years.

During a memorial meeting for T. H. Morgan in 1945, held in the Zoology Department, Dobzhansky described his arrival in the "fly room." "There were three people: Morgan sitting at his desk and two younger ones [presumably Sturtevant and Bridges] sitting on top of a desk. It was very informal." After introducing himself, Dobzhansky asked Morgan what kind of mathematical preparation he needed for his work. "Oh, the multiplication table," Morgan said. "Knowledge of long division is also useful."

Dobzhansky adjusted well to the Morgan laboratory, and when Morgan moved to Caltech (California Institute of Technology) in 1928, he invited Dobzhansky to come along. He stayed there until 1940, when he came to Columbia. When I first met Dobzhansky in 1943, he asked me what I thought was the most important goal for genetic research. I said, "The application of chemistry and physics to study the action of genes." I could see from his expression that he did not like my reply. For him, the basic unit was the gene, and there was no necessity to go below that to the level of chemistry. Later he sometimes referred to me disdainfully as the biochemist. On the other hand, he told me very proudly that once he had done a biochemical experiment using a Warburg manometer.

My laboratory was next to Dobzhansky's office, and as the doors were usually kept open, I could hear his loud and very distinctive voice. When he had a visitor, sometimes there were heated discussions, usually about the question of whether two related organisms belong to the same species or to different species. I must confess that I did not find these discussions very interesting.

Dobzhansky was a gifted linguist. He told me once that he spoke seventeen languages. Because of the many nationalities in New York, he felt that the only language he could speak safely in the subway without being understood by others was Southern Kirghizian. At one time, he was invited to give a series of lectures in Brazil. He did not know Portuguese, so he booked a passage on a ship to Rio de Janeiro, and during the voyage, he studied the language. On arrival, he delivered his lectures in Portuguese.

For my research project, I had to find a spectrophotometer in New York where I could measure my pigments. At Harvard, I had used an instrument that was very cumbersome because one had to sit in a dark room and match, at each wavelength, the light transmitted by the pigment solution against a known standard. Recently, a spectrophotometer had been developed in which light transmission could be measured electronically with a photoelectric cell. With this instrument, developed by Arnold Beckman and named for him, one could measure light absorption in the ultraviolet as well as the visible part of the spectrum. However, there was only one such instrument in New York City—in the laboratory of Dr. Leonor Michaelis at the Rockefeller Institute.

In looking for a way to have access to this instrument, I found out that Professor Pollister, a cytologist in the Zoology Department, was collaborating with Dr. Mirsky, a biochemist at the Rockefeller Institute, on studying the structure and chemistry of cell nuclei. I asked Professor Pollister if he could introduce me to Dr. Mirsky and tell him what I had in mind. In this way, I made contact with Dr. Mirsky, who invited me to his laboratory and took me to see Dr. Michaelis. The latter agreed to let me use his instrument. This was of great advantage for my work and made my life much easier.

I made many trips to the Rockefeller Institute to measure my pigment solutions, and most of the time, I visited Mirsky in his laboratory. He was very friendly, and we had many interesting conversations. He frequently took me to lunch where he introduced me to other members of the Rockefeller Institute. For me, it was a novel experience to meet these distinguished scientists informally over a lunch table.

Mirsky started to work at the Rockefeller Institute in 1927, where he continued previous studies on hemoglobin and other proteins. He investigated the phenomenon of protein denaturation in which protein molecules come apart and showed that this process was reversible. In 1936, he went for a year to Caltech to work with Linus Pauling on protein denaturation. Pauling had the idea that denaturation involved the breaking of hydrogen bonds within the protein molecules. Pauling, a physical chemist, had done pioneer work on the nature of such bonds.

In 1938, Mirsky started to study the chemical constituents of the nuclei of cells. Together with Pollister and Hans Ris, a Swiss cytologist, he did fundamental work in the field. He demonstrated that the DNA content of different organs of an animal or plant was the same but that the DNA content varied from species to species. He showed that the DNA content of eggs and sperm was half that of the other organs, as expected from half the number of chromosomes in these germ cells.

When I met Mirsky, he was actively engaged in these studies. He talked about his work with great enthusiasm. At one time, he showed me a test tube with DNA in solution. He added a little alcohol to the clear liquid and stirred it with a glass rod. He twirled this rod and then pulled it out. It dragged with it a layer of viscous, slimy material. "This is DNA," he explained. This was my first view of the genetic material.

One time, he invited me to come home with him and meet a friend. When we entered, there was a man lying on the sofa in the living room. Mirsky introduced me to him, but I didn't catch his name. "Tell him about your spectra of the eye pigments," he said. The man seemed to understand what I was talking about and even said that the light absorption curves in the ultraviolet part of the spectrum I showed him were to be expected. I was wondering how he would know what to expect, but then it became clear when I found out that his name was Linus Pauling.

In 1945, I had the opportunity to spend the summer at Cold Spring Harbor. This came about as follows. One day, Dobzhansky told me that he had rented a cottage at Cold Spring Harbor for the summer but that he had to go away and couldn't take it. Would I be interested to rent it? I was delighted by this offer. I spoke to my mother, and we decided to take it. There was a beach, and we could escape the heat of the summer in New York City. There were also laboratory facilities where I could work on the eye pigments.

An unexpected benefit of my stay at Cold Spring Harbor was the phage course (a course on bacterial viruses) taught that summer for the first time by Max Delbrück. Bacteriophages had become favorable material for the study of replication, and Delbrück was a pioneer in the new field of molecular biology. Unfortunately, I did not take the course, but I talked to the people taking it and to Max Delbrück. A few times, my mother prepared lunch for this group, and it was served on a long table in front of our cottage with Delbrück presiding at the head.

In the spring of 1946, I thought that I had enough experimental results to write my PhD thesis. I discussed it with Dobzhansky, and he approved of my plan. He then appointed a committee to serve as examiners for my thesis defense. It was a fairly large group, about six or seven, but I only remember, besides Dobzhansky, Charles Dawson, Alfred Mirsky, and Selig Hecht. I wrote my thesis in about a month, and a few weeks later, I had my formal examination, which I passed without difficulty.

In retrospect, looking at the work I did, although it seemed to satisfy the examiners, I find it appalling. Not only did it fail to contribute anything of significance to the general knowledge of gene action but even my attempts to determine the chemical nature of the pigment were also unsuccessful. In my largest preparation, I had raised sixty thousand flies and obtained 3.9 milligrams of pure pigment. In my hands, this was not enough material for a chemical analysis. Like Beadle previously, I learned the hard way that *Drosophila* eye pigments were not suitable material for biochemical genetics.

For postdoctoral work, I was eager to go to Caltech, where Beadle had become chairman of the Biology Division and had taken his coworkers with him when he moved there from Stanford. Dobzhansky told me that there were two new fellowships available in Beadle's department and suggested that I should apply. There was no formal application, and I just wrote a letter about my education and PhD research. A few weeks later, Dobzhansky handed me a little piece of paper, which said, "Maas accepted. Sturt" (Sturt is an abbreviation of Sturtevant). This was amazing, both for its informality and simplicity. From the viewpoint of the present time, it is incredible. I was elated to be able to go to the mecca of biochemical genetics.

I arrived at Caltech in November 1946, after a three-day train journey from gray and cold New York. The two trains I took (I changed trains in Chicago) had no sleeping cars, but the coach seats were very comfortable. I was absorbed in the panoramic views from the train window by the passing landscapes, the endless plains of the Midwest, the Rocky Mountains, the colorful deserts of New Mexico and Arizona, and finally, the lush orange groves and palm trees in mild and sunny California. It was a novel and exciting experience.

The campus of Caltech was spacious and secluded, with Spanish-style buildings separated by lawns and flower beds. A passageway lined with olive trees led across the campus, ending in front of the stately faculty club building called the Atheneum. The building housing the Biology Department, called Kerckhoff, was located at the other end of the campus. There were three long floors in this building with wide corridors. Beadle's office and the laboratories of people working with *Neurospora* were on the second floor. Sturtevant and his associate Ed Lewis were on the third floor. I was assigned a laboratory on the second floor.

I rented a furnished room in the house of a charming English lady, Agnes Thompson, within walking distance of the campus. It had a garden with an avocado tree and a fig tree. Soon I found out one of the distinguishing features of Pasadena. When I walked to work, I was always the only person walking on the street. People seemed to go places only by car. I did not think that it was necessary for me to have a car, but I found out that public transportation was very scanty, and I was practically confined to the area around Caltech. After a year, I finally broke down and bought a car. It was a 1928 Model A Ford Sedan, which I acquired for fifty dollars. Now I could explore the many interesting places in Southern California.

For evening meals, a group of five postdoctoral fellows, including myself, joined in their preparation in the "kitchen" used for sterilizing *Neurospora* culture media and washing laboratory glassware. We took turns cooking with the restriction that one could not spend more than forty cents per person. It was a congenial arrangement, although the meals were bizarre. An exception was a meal prepared by Hans Gloor, a postdoctoral fellow from Switzerland. It was an excellent steak dinner, but most of the cost for this meal had come out of Hans's pocket.

The fellowship I had received was called a Gosney Fellowship, named after a banker in Pasadena who had endorsed it. It carried a generous stipend of $2,500 per year and left me complete freedom in the choice of a research project. The daughter of Mr. Gosney, Lois Gosney, was interested in the welfare of the two Gosney fellows (the other was Ray Owen, a geneticist from the University of Wisconsin who had discovered the important phenomenon of immunological tolerance in twin cattle) and invited me on several occasions, the last one being the opening of the Palomar Observatory near San Diego in 1948. On the way back to Pasadena, we passed a large resort hotel in the desert. Lois turned to me and said, "This is a great place, and the best part is, they don't take Jews." After I told her I was Jewish, she was clearly flustered and embarrassed, but in the end, she qualified her statement, and we remained friends.

The freedom of choice put me into a quandary. I wanted to work on *Neurospora* and learn the technique of handling this organism. My main interest was to elucidate the nature of the genetic control implied in the one gene—one enzyme hypothesis: Does the gene determine the structure of an enzyme protein, or does it merely govern the rate of its formation? Unfortunately, this key question was not studied directly by any of the members of Beadle's group. Most of them had settled down to more or less routine projects dealing with the use of mutants to work out the biochemical pathways in the formation of metabolites. This was very important for biochemistry, but it did not address my question.

Beadle at this time had given up his own laboratory and was engaged heavily in his administrative duties as chairman of the department. I think that he was aware of the direction *Neurospora* research had taken. One time, as I was walking next to him along the corridor, he said to me, "At Stanford, we worked in a basement under crowded condition, and we produced great work. Now that we have all the space and facilities one could hope for, we will probably never do a damn thing."

I did not see much of Beadle, but I was strongly impressed by his simple and direct manner. There was nothing of the Europeans' "Herr Professor" about him. One morning, I arrived early, and passing by his office, I saw him vigorously sweeping the floor. I stopped, and seeing my astonished face, he said, "It doesn't matter what you do as long as you do it well." My favorite quotation of Beadle is what he said in answer to a question about the role of the double-helix structure of DNA in replication: "Do not discard a hypothesis just because it is simple—it might be right."

Of the people in the *Neurospora* group who had come together with Beadle from Stanford, the one I talked to most often was Norman Horowitz. Norm, to his friends, had a sharp and wide-ranging mind masked by a relaxed and easygoing manner. He had been a graduate student at Caltech in Morgan's department and had received his PhD in 1939. After that, he went on a National Research Council Fellowship to Stanford, and in 1942, he joined Beadle's group. He had a deep interest in evolution and in the origin of life and in 1945 published a very interesting paper on the evolution of biosynthetic pathways. In 1946, he was working on the biosynthesis of the amino acid methionine, but he clearly had a broad interest going beyond one specific pathway. In looking for a suitable system in *Neurospora* to study the question I was posing, I came across some early papers that described the production of peptidases. These enzymes break down peptides consisting of a small number of amino acids to the level of the constituent amino acids. Since a given peptidase can attack several peptides, it occurred to me that one could look for mutants in a peptidase in which the pattern of peptides attacked by the peptidase was changed. This would imply that the mutation had changed the structure of the enzyme protein. I discussed my idea with Norm, and he thought it was reasonable.

I started to work on my project and soon ran into a serious difficulty. The peptides I needed were not available commercially, and I had to synthesize them. Such syntheses are difficult, and I was not proficient in doing this kind of chemistry. My situation was similar to what it had been at Columbia, where there was nobody in the department who was working in the area of my problem and I was left on my own. I kept on struggling with my problem, but at the end of my two-year stay, I had not gotten very far. At that point, I realized that another kind of mutant, called temperature-sensitive mutant, would have been far better material for my problem. Such mutants had been found in *Neurospora*. They could not grow at the upper range of the normal temperature of growth without a required growth factor but could do so at a lower temperature. If one could extract the enzyme affected by the mutation, one could test to see if it became more sensitive to heat inactivation in the mutant. Unfortunately, it was now too late to start another research project.

In discussing my failure with Norm, he tried to console me by saying, "Sometimes it is difficult to attack a problem directly. You just have to wait until nature turns her back." It was a prophetic statement.

During the following years, Norm became the chief spokesman for the one gene—one enzyme hypothesis. At the 1951 Cold Spring Harbor Symposium on "Genes and Mutations" in which the validity of the one gene—one enzyme was seriously questioned by several speakers, Norm reported experimental results (to be described in the next chapter) that supported the hypothesis and defended it against the criticisms leveled against it. A few years later, after the known structure of DNA made it possible to envisage how the information contained in the sequence of nucleic acid bases could be transcribed

into the sequence of amino acids of proteins, the credibility of the one
gene—one enzyme hypothesis was finally established.

In the 1960s, Norm became involved in planetary science, then
carried out in the Jet Propulsion Laboratory in Pasadena. Between
1965 and 1970, he was chief of its bioscience section. He was largely
responsible for the design of the experiments carried out by the Viking
lander to test for the possibility of life on the surface of Mars. At that
time, he sent me a picture of the lander on Mars, together with a note
saying, "There is no life on Mars."

My last encounter with Norm occurred in 2000 when I came to
Pasadena to hear what he had to say about the manuscript of my book
on gene action I had sent to him for his comments. He had read it
with great care and made many helpful suggestions. During my visit,
we attended a seminar given by Richard Axel on his work on the
genetics of smell receptors for which he subsequently was awarded the
Nobel Prize. As we were walking out, Norm turned to me and said, in
his characteristic deadpan fashion, "Now, we have to propose the one
gene—one smell hypothesis."

Norm died in 2005. I attended a memorial meeting at Caltech a
few months after his death. It was a moving occasion that brought out
the mentality of Norm as described by Robert Metzenberg, a former
Caltech graduate student, in an obituary: "The blunt and ever-civil
truth teller." The only survivors of the 1946 Caltech group, Ray Owen
and myself, were present at this meeting.

In the fall of 1948, I returned "back east" to start a new job at the
Naval Medical Research Institute in Bethesda. I had previously been
interviewed for this job, and for this purpose, I had been flown on navy
transport planes from Los Angeles via San Francisco to Washington. It
was my first airplane trip and was very exciting.

During the spring of 1947, I had also made a trip "back east,"
to attend the Cold Spring Harbor Symposium on "Nucleic Acids
and Nucleoproteins" (fig. 2). For this occasion, I had gotten a lift to
Madison, Wisconsin, with Ray Owen, who went there with his wife
and two children to visit other family members. From Madison, I went
by train to New York. At the Cold Spring Harbor Symposium, I met
Joshua Lederberg, who, as a medical student at Columbia, had done
some experiments on *Neurospora* in Francis Ryan's laboratory. He
subsequently conceived the idea of looking for a gene-transfer system
in bacteria, and he went to Tatum's laboratory, who was at that time
in Yale, to test his idea with mutants of *E. coli*. He was successful in
demonstrating genetic transfer in a cross between two strains, similar to

sexual reproduction in higher forms. For his pioneering work, he was awarded the Nobel Prize in 1958 together with Beadle and Tatum.

In 1948, I was glad to return to the East Coast. I had many interesting and enjoyable experiences in Pasadena, but I never felt quite at home. I missed the change of seasons and the green forests and meadows. I also missed the busy street life of New York. I had sold my 1928 Ford and bought a 1933 Plymouth two-seater coupé for $165, which I used for the return trip to New York. A graduate student of Norm, Bernie Strauss accompanied me. The trip went well until it started to rain after we had crossed the Rocky Mountains. As it never rained in California, I had not bothered to check the windshield wipers, and they did not work. I managed to reach a gas station by staying close behind a truck until I could get the wipers repaired. A more serious difficulty arose when we reached the Pennsylvania turnpike. There was a loud bang, and then the motor went dead. It turned out that the piston had gotten lose and had gone through the engine case. The car was towed to a garage, and the mechanic thought he could fix it, but it would take several days. Bernie decided to continue his trip by bus, and I stayed for two days until, luckily, the car was repaired. I arrived in front of my mother's apartment on Central Park West without further incident.

Chapter 3

Genetic Control of Enzyme Formation in *E. coli*

Figure 1. Members of the Davis laboratory in 1950. Standing from left to right: Gordon Allen, Bernard Davis, the author, Henry Vogel. Sitting, at left, Elizabeth Mingioli, at right, Molly Sanderson.

I set up my laboratory at the Naval Medical Research Institute with the view of studying temperature-sensitive mutants of *Neurospora*. The facilities were good, and I was free to work on a problem of my choosing. The only shortcoming was that there was nobody there who was working in my field of interest, the closest being one geneticist who was studying the lethal effects of radiation in mice. Scientifically, I felt isolated.

After a few months in Bethesda, I was contacted by Bernard Davis, a medical research scientist who had recently discovered a method for the efficient isolation of mutants in the bacterium *Escherichia coli* (*E. coli*). He had set up his own laboratory at Cornell University Medical College in New York City, and he was looking for a genetically oriented collaborator to carry out biochemical studies similar to those done with *Neurospora*. This sounded enticing, and I made an appointment with him. I met him in his laboratory, and over a three-hour lunch at a nearby Czech restaurant, we talked about our interests and scientific goals. I was very impressed by his personality, and I think the feeling was mutual. The outcome was that I should ask my employers to send me "on loan" to his laboratory for an indefinite period. After returning to Bethesda, I had no problem to convince the director of the institute that it would be beneficial to spend some time in Davis's laboratory.

When I moved to Bernard Davis's (Bernie to his acquaintances) laboratory, I did not expect that the indefinite period would turn out to be seven years—first three years at Cornell University Medical College and then, after a year's absence in Boston, three years at New York University School of Medicine. Before I describe my experiences during these years, I shall give a brief account of Bernie's life that led up to, what was for me, a fateful meeting.

Bernie was born in 1916 in Franklin, Massachusetts, and was the son of Jewish immigrants from Lithuania. In his family, there was great emphasis on learning and intellectual achievement. All four children graduated as valedictorians from the local high school. Despite the limited financial resources generated by his dry goods store, Bernie's father managed to provide his children with an education at Harvard (for the two sons) and Radcliffe (for the two daughters).

At Harvard, Bernie concentrated on the hard sciences with emphasis on biochemistry. After graduation, he vacillated between graduate work in chemistry and medical school. For practical reasons, he decided for the latter and enrolled at Harvard Medical School. During the medical curriculum, he also worked in the laboratory of E. J. Cohn, a pioneer in protein chemistry. He graduated in 1940 with the very rare degree of MD, summa cum laude.

In 1942, he began a research career as a commissioned officer in
the US Public Health Service. He was assigned to work on serological
tests for syphilis. To prepare himself for this task, he spent two years
as an apprentice in two laboratories that carried out research in
immunochemistry, first with Elvin Kabat at Columbia University and
then with Jules Freund at the Public Health Research Institute of
the City of New York. In 1945, the US Public Health Service offered
Bernie his own laboratory to work on basic science problems related
to tuberculosis. He prepared himself for this by spending two years in
the laboratory of René Dubos at the Rockefeller Institute. During this
time, he contracted tuberculosis and had to undergo surgery, followed
by a protracted recovery period. It was during this period of reading
and reflection that he formulated the plans for his future research.
A deciding factor was a review by Beadle on the use of nutritionally
deficient mutants of *Neurospora* for biochemical studies. This review
planted a seed in Bernie's mind, which germinated during a seminar
when the speaker noted that penicillin kills bacteria only when they are
growing. He realized that in a culture of *E. coli* growing in a "minimal"
medium, penicillin should kill off the normal cells while allowing rare
mutants with additional growth requirements and therefore unable
to grow to survive. Thus was born the penicillin method for the
isolation of nutritionally deficient mutants blocked in the synthesis of
essential metabolites, analogous to the mutants isolated in *Neurospora*.
The method turned out to be very efficient, and at the time I joined
his laboratory, he had accumulated a large number of mutants with
different growth requirements (fig. 1).

After several months in Bernie's laboratory, I decided to stay
there permanently. In June 1949, I changed employers and became,
like Bernie, a commissioned officer in the US Public Health Service.
During this time, the emphasis of Bernie's research was on working out
biochemical pathways, especially the steps in a common pathway for
the synthesis of the aromatic amino acids tyrosine, phenylalanine, and
tryptophan. To carry out this kind of analysis, it is necessary to identify
chemically the intermediates of a pathway. For this purpose, Bernie
acquired several associates with training in organic chemistry. Besides
the aromatic pathway, the formation of the amino acids proline, lysine,
and methionine were investigated with the use of mutants. These
studies established Bernie as a widely recognized figure in microbial
physiology.

The work on biosynthetic pathways constituted the
bread-and-butter research of the laboratory. There were other studies
with mutants that were not as well-defined as biosynthetic pathways, but

they were of greater general interest in foreshadowing developments in molecular genetics. Such studies included my own work with a temperature-sensitive mutant that demonstrated that a mutation could alter the structure of an enzyme. The latter came about as follows.

I had told Bernie about my idea of using temperature-sensitive mutants to investigate the role of genes in the control of enzyme formation. He was supportive of my plans to carry out such studies with mutants of *E. coli*. There was only one temperature-sensitive mutant in his collection that required the amino acid serine for growth at 37°C but not at 30°C. Although the enzymatic reaction affected by the mutation was not known, I decided to work with this mutant. I used DL-serine as my source of serine, and to my surprise, I discovered that D-serine, the unnatural form of the amino acid (the natural form is L-serine) inhibits the growth of *E. coli*. This interesting finding led to an investigation of the mechanism of D-serine inhibition, and I found that D-serine inhibits the formation of the vitamin pantothenic acid. In this case, I could obtain the enzyme that was inhibited by D-serine in extracts. Serendipitously, I subsequently obtained a temperature-sensitive pantothenic acid—requiring mutant affecting this enzyme, and I could now study the question I had asked originally. I found that the enzyme of the mutant was much more heat-labile than the corresponding enzyme from the normal strain. This led me to conclude that genes control the structure of enzyme molecules rather than the rate at which they are processed. I shall briefly describe how I found this mutant because I thought that nature rewarded me for being conscientious. I had previously obtained an absolute pantothenic acid—requiring mutant that produced no pantothenic acid—synthesizing enzyme at any temperature and which I used in bioassays to measure the amount of pantothenic acid produced in enzymatic reactions. In those assays, I always had a set of tubes with known amounts of pantothenic acid as standards, including a zero tube without pantothenic acid. Usually I discarded these tubes at the end of the day, but one time, I forgot and left the rack with tubes on the bench. The next morning, I noticed that the zero tube had grown out. I expected a reversion to wild type, and to test this, I plated a sample on minimal medium and incubated the plate at 37°C as usual. Surprisingly, nothing grew out. Then it dawned on me that the reversion could have been not to the wild type state but to a temperature-sensitive state because the tubes had been left at room temperature. Sure enough, when I incubated the minimal agar plate at room temperature, the cells grew. Nature had kindly shown me how to do the experiment.

There was a problem that I had to consider before drawing my final conclusion. There was the probability that the mutation to temperature sensitivity was due to the production of a substance that made the *E. oli* more heat labile. At the time, the physicist turned biologist, Leo Szilard, visited our laboratory, and I told him about my problem. "There is a simple experiment to test for this possibility," he said. "Study the temperature effect in a mixture of the mutant and the normal enzyme. With an inactivating substance, both enzymes should become more heat-labile. With the mutant enzyme being an altered protein, you should get two distinct curves." I found the latter in my experiment. This was the first of frequent subsequent events when I received valuable advice from this unusual man who delighted in talking to people whose research interested him and in giving them advice.

In the previously mentioned 1951 Cold Spring Harbor Symposium, Norman Horowitz reported results obtained with temperature-sensitive mutants of *Neurospora* and of *E. coli* that refuted an objection raised by Max Delbrück against the one gene—one enzyme hypothesis. Delbrück had maintained that the mutants isolated at Caltech on the basis of growth factor requirements were a small fraction of all possible mutants. Norm showed that among a large number of mutants isolated solely on the basis of temperature sensitivity, a significant proportion could grow on the standard complete medium at the higher temperature and were therefore of the nutritionally-deficient type.

Following Norm's talk, I presented my results with the heat-labile enzyme in the pantothenic acid—requiring mutant. These results predicted the picture of gene action that emerged in subsequent years after the discovery of the double-helix structure of DNA in 1953.

In 1952, I left the Tuberculosis Research Laboratory to spend a year in Fritz Lipmann's laboratory in Boston, as will be described in the next chapter.

After I returned from Boston at the end of 1953, Bernie was in the process of dissolving the Tuberculosis Research Laboratory. He had decided to accept a position as chairman of the Pharmacology Department at New York University School of Medicine. He offered me a position as assistant professor, which I accepted. My experiences there will be described in chapter 5.

Chapter 4

A Sojourn in Fritz Lipmann's Laboratory

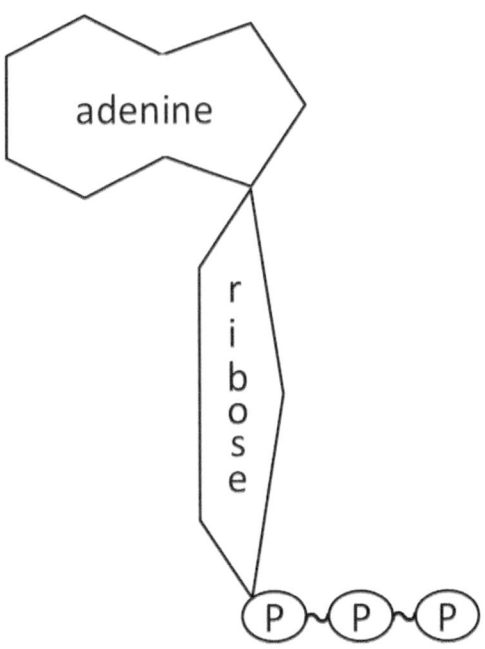

Figure 1. Schematic presentation of the chemical
structure of adenosine triphosphate (ATP).

Figure 2. Members of the Lipmann laboratory in 1953. Lipmann is in front row, center. To his left: Mary Ellen Jones, David Novelli. At right corner, John Gregory. The author is behind Lipmann to the right.

In 1950, I attended a lecture given by Fritz Lipmann in New York. He talked about a coenzyme he had recently discovered. (A coenzyme is a substance required by certain enzymes for activity.) The coenzyme Lipmann had discovered was involved in the enzymatic acetylation of a number of metabolites and, for this reason, was named coenzyme A. What aroused my interest was that the major constituent of coenzyme A is pantothenic acid. I saw an opportunity that by working in Lipmann's lab, I could learn about the mechanism by which my enzyme catalyzed the coupling of pantoic acid with beta alanine to form pantothenic acid. I contacted Lipmann, and to my delight, he invited me to come to Boston in the fall of 1952 and spend a year in his laboratory.

During the interval between 1950 and 1952, I concentrated my efforts on pantothenate synthase (as I named my enzyme). For activity, the enzyme required adenosine triphosphate (ATP, fig. 1) as a source of energy and monovalent cations (K^+ or NH^+) and divalent cations (Mg^{++} or Mn^{++}) as activators. I did quantitative studies on these requirements as well as on other physical and chemical properties of the enzyme, and I published these studies in the *Journal of Biological Chemistry* before leaving for Boston.

During this period, I also studied, together with my colleague Henry Vogel, the biosynthesis of pantoic acid. We found that this substance was formed from ketovaline, which is a precursor of the amino acid valine. Further studies carried out on the biosynthesis of beta alanine, the other precursor of pantothenic acid, showed that it could arise from either aspartic acid or glutamic acid. After finishing these experiments, I went to Boston in August 1952.

Lipmann's laboratory occupied the whole fourth floor of a new research building at the Massachusetts General Hospital. There were three staff members, Mary Ellen Jones, David Novelli, and John Gregory (fig. 2), and eight temporary researchers, including the German biochemist Feodor Lynen, who, like Lipmann, later won a Nobel Prize.

To my surprise, Lipmann's interest in my enzyme was not in its role in coenzyme A formation but as a model for peptide bond formation in protein synthesis. He had become involved in this subject and had close contacts with the group of Paul Zamecnik located on the floor above, who were developing an in vitro system for studying protein synthesis.

There was an interruption in my getting started to work on the mechanism of my enzyme's action. Shortly after my arrival, Henry Vogel came to Boston and talked to Lipmann and me about his recent studies on the biosynthesis of the amino acid arginine. He mentioned that the first step in this pathway was the conversion of glutamic acid to acetylglutamic acid. Lipmann became interested because of the possible involvement of coenzyme A in the mechanism. After Henry left, he asked me to test for this possibility. I carried out this enzymatic reaction in extracts of *E. coli*, and it did demonstrate a requirement for coenzyme A. Within a short time, we published my results, and I felt proud to have a joint paper with Lipmann.

I finally started to work on pantothenate synthase together with Dave Novelli, and my first finding was that it was not the terminal bond but the subterminal bond of ATP that was used as an energy source, resulting in the formation of pyrophosphate (PP) and adenylic acid (AMP). This was unusual; so far, it had been found that the terminal bond was used, resulting in the formation of P and ADP.

Lipmann told Lynen about our results, and he suggested that to further elucidate the mechanism of this reaction, we should carry out exchange reactions between radioactive PP (PP^{32}) or radioactive AMP (AMP^{32}) and ATP. We did this and found exchange between both AMP^{32} and ATP and PP^{32} and ATP, but the latter required the presence of pantoate. On the basis of these results, Lipmann formulated the following mechanism of the reactions, which had the unusual feature that the enzyme (E) itself formed a high-energy bond with PP^{32} to produce an intermediate component of the reaction, as shown in the following scheme:

$$E + ATP \leftrightarrow E\text{-}PP + AMP$$
$$E\text{-}PP + pantoate \leftrightarrow E\text{-}pantoate + PP$$
$$E\text{-}pantoate + beta\ alanine \leftrightarrow pantothenic\ acid + E$$

Lipmann was quite excited about this novel mechanism and presented it the following year at a symposium on "The Mechanism of Enzyme Action" at Johns Hopkins University. Unfortunately, as we shall see in the next chapter, it was all wrong.

In the fall of 1953, I returned to New York with the feeling of having spent a fruitful year in becoming acquainted with the world of biochemistry. In 1957, Lipmann moved to the Rockefeller Institute in New York, and I had many contacts with him and his wife, Freda. Whenever I saw Lipmann, I enjoyed talking to him about biochemistry. I had the feeling of being on the same wavelength with him.

Chapter 5

Continuing Pantothenic Acid Synthesis in New York

A few months after my return to Bernie's laboratory, we moved to the Pharmacology Department in the new Medical Science Building of New York University. In the interim, I had continued with the purification of pantothenate synthase and, to my surprise, found that during this purification the exchange reaction between AMP^{32} and ATP disappeared. Apparently, it was due to another enzyme.

After setting up my own laboratory at New York University, I was joined by two associates—Michael Yarmolinsky, a postdoctoral fellow, and David Hogness, on his way from Jacques Monod's lab in Paris to start a new position in Arthur Kornberg's department at Washington University in St. Louis. They joined me in working out the mechanism of the pantothenate-synthesis reaction. I also had a technician, Rita Altszuler.

On the basis of having only one exchange reaction with PP^{32} and ATP and requiring pantoate, I proposed the following model:

ATP + pantoate $\leftrightarrow$ pantoyl-AMP + PP
Pantoyl-AMP + beta alanine $\leftrightarrow$ pantothenate + AMP

In this scheme, the enzyme has only a catalytic function.

We found evidence for this scheme from two kinds of experiments: (1) In the presence of enzyme, ATP, pantoate, and hydroxylamine, there is an equimolar formation of pantoyl-hydroxamic acid and PP, demonstrating the formation of a pantoyl compound as an intermediate and (2) that the pantoyl compound in pantoyl-AMP was shown in experiments with pantoate labeled with O^{18}, which was prepared for us by Daniel Koshland at the Brookhaven National Laboratory. In the pantoate molecule, there are two oxygen atoms, and both were labeled. In the formation of pantoyl-AMP, both oxygen atoms are retained. In the subsequent formation of pantothenate, one of the two oxygen atoms is expected to be transferred to AMP, and this is what we found.

An organic chemist in our department, Felix Leitner, tried to obtain even more direct evidence for the proposed mechanism by synthesizing pantoyl-AMP and testing it as a precursor for pantothenic acid. Unfortunately, all his attempts failed because of the easy conversion of pantoate to pantoyl lactone.

After I told Lipmann about our results, he mentioned them to Mahlon Hoagland who was working in Zamecnik's laboratory on the activation of amino acids in protein synthesis. Mahlon found the same mechanism with the formation of amino acid-AMP compounds as intermediates.

The lesson I learned from my experiments after returning from Boston was expressed by F. Racker in the following dictum: "Do not waste pure thought on impure enzymes."

I presented our studies on pantothenic acid synthesis at two international congresses of biochemistry in Brussels in 1955 and in Vienna in 1958.

Chapter 6

Feedback Control of Enzyme Formation in Arginine Biosynthesis

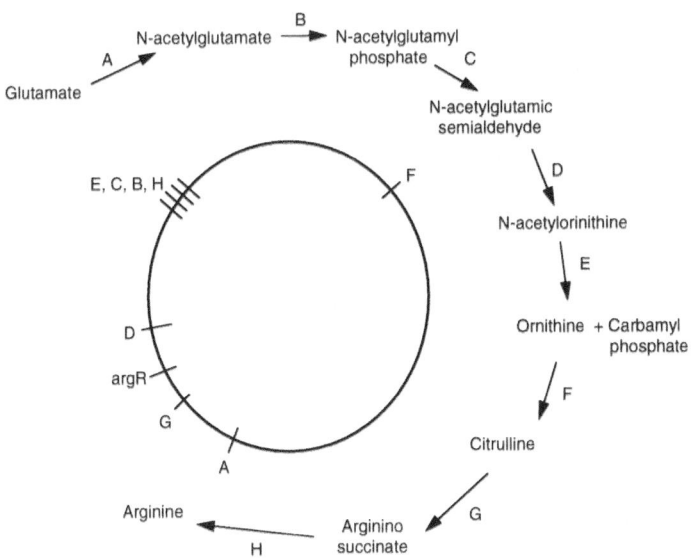

Figure 1. The enzymatic steps in the biosynthesis of arginine. The map position of the genes controlling these steps is shown in the circular map of the E. coli chromosome. The regulatory gene ArgR is also shown.

In 1955, after having completed my work on pantothenic acid, I returned to the original question I had asked about gene action. Having demonstrated to my satisfaction that genes control the structure of protein molecules, I now asked the second question: What controls the rate of gene expression?

I used the same approach I had used previously: to look for mutants in which the rate of formation of the corresponding enzyme is altered. As a suitable test system, I chose the formation of ornithine transcarbamylase (OTCase) that catalyses the conversion of ornithine to citrulline in the synthesis of arginine (fig. 1). I looked for cold-sensitive mutants because I thought it was likely that in such mutants, the rate of enzyme synthesis would be altered rather than the structure of the enzyme. Starting with a completely blocked OTCase mutant, I obtained a cold-sensitive mutant that grew at 37°C without arginine but required arginine for growth at 25°C. OTCase activity was easy to measure in extracts. As expected, the enzyme was produced after growth at 37°C in medium without arginine, but not after growth at 25°C in medium with the required arginine. A surprise came when, as an extra control, I grew the cells at 37°C with arginine: no OTCase was produced. Subsequently, I found inhibition of OTCase formation in the wild-type parent strain as well, both at 37°C and at 25°C. In this roundabout way, I had discovered that feedback inhibition of OTCase formation by arginine is a general regulatory phenomenon.

Unknown to me, Henry Vogel, at the time at Yale University, had demonstrated inhibition by arginine of acetylornithinase, the enzyme that converts N-acetylornithine to ornithase. In a publication in 1957, he coined the term *enzyme repression* for this type of feedback control. He named a substance-inhibiting-enzyme synthesis a repressor. Eventually, repression by arginine for all the enzymes of its biosynthetic pathway was demonstrated. The same kind of control mechanism was found in the biosynthetic pathways of other amino acids.

After discovering that added arginine repressed enzyme formation, I wanted to know if internally produced arginine could have the same effect. I made an observation that suggested that this was indeed the case. A culture grown with arginine was transferred to a medium without arginine, and growth was continued. There was an initial burst of OTCase synthesis for about twenty minutes followed by a sharp decrease in the rate. This result was very different from the results obtained by Jacques Monod on the kinetics of betagalactosidase synthesis. He had found that this enzyme, after addition of the inducer lactose, was produced at a constant high rate for several hours. To explain my results, I postulated that during the initial burst, arginine would be produced rapidly inside the cells and would inhibit further OTCase synthesis. The problem was to perform an experiment that would test this explanation. To do this, I had to find conditions of growth under which arginine would not accumulate.

Luckily, I learned that such conditions could be obtained by growing an arginine-requiring mutant in a device for continuous growth called a chemostat, under conditions of the growth rate being limited by the supply of arginine. Under these conditions, the concentration of internal arginine was greatly reduced.

The chemostat was invented by Leo Szilard in 1950 at the University of Chicago where, together with Aaron Novick, he had set up a laboratory for the study of bacterial genetics and physiology.

At this time, in 1956, I was joined by a visiting scientist, Luigi Gorini. Together, we carried out experiments with a chemostat under conditions of arginine-limited growth. We found that OTCase was produced at a high constant rate, as expected from Monod kinetics. The enzyme was produced about five hundred times as fast as in a wild-type strain growing in a culture medium without arginine. We concluded that the bacteria have a very high potential for making OTCase but that the formation of the enzyme is throttled down by internally produced arginine to a low level yet sufficient for normal growth. Later, this kind of reserve potential was demonstrated for other enzymes of arginine biosynthesis. It was recognized as a general principle that feedback control using repression by the end product maintains enzymes at an adequate level for growth and that there is extra reserve capacity for enzyme synthesis, presumably to deal with emergencies.

The strain of *E. coli* we had used for the chemostat experiments was the W strain, named after Selman Waksman, the discoverer of streptomycin. When we tested the commonly used *E. coli* B strain for repression by arginine, we found a surprisingly different situation. There was the same low level of OTCase in the absence of arginine, but adding arginine caused an increase rather than repression of OTCase activity. Luigi Gorini took up the study of the puzzling difference and continued it after moving, together with Bernie Davis, to Harvard Medical School. He died in 1976, and I continued to work on the problem. A clear explanation was not obtained until 1994, as I shall describe subsequently.

An unexpected consequence of the chemostat experiments occurred in 1957 when I attended the Federation Meetings in Chicago. I stayed at the Quadrangle Club, which Szilard was also staying in. I met him at breakfast and told him that the mutant I had used had a genetic block early in arginine biosynthesis and did not produce ornithine, the substrate of the OTCase reaction. Yet OTCase could be produced at a very high rate. This was contrary to the idea of Monod that in the case of β-galactosidase, induction of the enzyme to produce large amounts required the presence of the substrate lactose, or an analog of the substrate. On the basis of a unitary hypothesis, induction of β-galactosidase could be due to removal of repression. The inducer acted secondarily in removing a repressor.

Szilard was impressed by my idea, and during discussions during breakfast on the following days, we became convinced that it could be true.

Early in 1958, he presented my hypothesis at a seminar at the Pasteur Institute in Paris. At that time, there was an experiment in progress in Monod's laboratory that was designed to test Monod's hypothesis. It was being carried out by Arthur Pardee, a recent arrival from the University of California at Berkeley. The idea was that according to Monod's scheme, in zygotes formed in matings between an inducible wild-type Lac⁺ strain and a constitutive Lac⁺ mutant, the phenotype of the mutant, producing an internal inducer, should be dominant. The result of the experiment was the opposite: in the phenotype of the Lac⁺ wild type, no enzyme formation in the absence of an inducer was dominant. This demonstrated the presence of a repressor, which was inactivated by the added inducer. This classical experiment came to be referred to as the Pajamo (Pardee, Jacob, and Monod) experiment. What Monod had called an inducer became now an antirepressor.

In zygotes formed in matings, the diploid state is only temporary. Later, Edward Adelberg and François Jacob constructed permanent diploid strains, inserting a plasmid carrying the Lac genes into a strain carrying the Lac genes in the chromosome. This confirmed the dominance of the wild type Lac⁺ phenotype.

Chapter 7

The Genetic Basis of Enzyme Repression

It seemed unlikely that arginine by itself could bring about the repression of the eight enzymes in its biosynthesis. I therefore expected that there is another substance that acts together with arginine to bring about enzyme repression. One possibility of finding this substance was to look for mutants that were no longer repressible by arginine.

In the case of several amino acids, there were chemical analogs known that inhibited growth by competing with the amino acid in protein synthesis. One could look for mutants that are resistant to the growth-inhibiting effects of such analogs. This might be due to the loss of repression; that would result in overproduction of the competing amino acid.

In the case of arginine, such an analog is canavanine, a substance produced by canava beans. Inhibition of growth by canavanine is overcome by the addition of arginine.

I started to isolate canavanine-resistant mutants of *E. coli* strain W in 1957. At that time, my coworkers were James Schwartz, a medical student, and Eric Simon, a biochemist. All the mutants we isolated were still repressible by arginine and found to be defective in a specific uptake and concentrating mechanism (permease) for basic amino acids, including canavanine. These findings gave my research a new direction, but they did not help our search for an arginine repressor.

During the summer of 1959, while I was a visitor in François Jacob's laboratory at the Pasteur Institute in Paris, I again looked for canavanine-resistant mutants, this time in the *E. coli* strain K12. As before, I used OTCase formation as an indicator of repressibility. All the mutants I isolated were derepressed for OTCase, producing large amounts of the enzyme. During the following years, they were shown to be derepressed for all the enzymes of arginine biosynthesis.

After my return to New York, I mapped these mutants in the *E. coli* chromosome. They all were located in the same site, presumably in the same gene. I named the gene ArgR (*R* for regulator). I proposed that ArgR controls the production of an "aporepressor," arginine being the "corepressor." The overall conclusion of these studies was that there are actually two kinds of genes: one kind determining the structure of enzyme molecules, the other producing a substance that regulates the rate of enzyme formation.

In 1959, mutants in each of the eight steps of arginine biosynthesis were available, and the genes controlling the production of the corresponding enzymes were mapped in our laboratory and in Gorini's laboratory. They were located in five sites in the circular chromosome with none of them overlapping the site of ArgR (fig. 1, chapter 6). The repressor, whatever its chemical nature, had to act at each of the five sites to regulate arginine synthesis.

Chapter 8

Feedback Inhibition of the First Enzyme in Arginine Biosynthesis

At the time of my discovery of repression of enzyme formation in 1956, a second kind of feedback regulation, inhibition of the activity of the first enzyme in a biosynthetic pathway, was discovered in isoleucine biosynthesis of *E. coli* by E. Umbarger and in pyrimidine biosynthesis of *E. coli* by R. A. Yates and A. Pardee. In subsequent years, as in the case of enzyme repression, this mechanism was demonstrated in other biosynthetic pathways.

For arginine, there were some experimental difficulties that had to be overcome before I could demonstrate, together with my associate A. Vyas, feedback inhibition by arginine of acetylglutamate synthetase, the first enzyme of arginine biosynthesis. Extracts of *E. coli* always yielded low activities of the enzyme, and the assay for acetylglutamate, which I had used previously in Lipmann's laboratory, was not very specific. We overcame these difficulties in 1962 by using a mutant blocked between glutamate and acetylglutamate as an assay organism and using a mutant blocked in the step after acetylglutamate for measuring enzyme activity in resting cell suspensions. Then we could demonstrate feedback initiation of acetylglutamate synthesis unambiguously. We also demonstrated repression of this enzyme.

The two feedback mechanisms complement each other. Repression keeps all the enzymes constantly at a low level but sufficient for the arginine requirement for growth. The high potential for enzyme synthesis provides a safety valve in case rapid enzyme synthesis is necessary for survival. Otherwise, repression saves energy by preventing unnecessary protein synthesis. Feedback inhibition of enzyme activity, which is created by adding arginine, is more flexible than repression, being quickly turned on or turned off. It is thus suitable for response to rapidly changing environments. It acts directly on the enzyme, whereas repression acts on the gene producing the enzyme, as we shall see in the following chapter.

Feedback inhibition of the first enzyme in the arginine pathway later had unexpected consequences. In 1969, I was investigating the possibility that the corepressor was not arginine itself but an arginine-tRNA compound. To study this, I was looking for canavanine-resistant mutants that have an altered arginine-tRNA. Among these, I found one whose growth was inhibited by arginine. Yet the arginine-tRNA in this mutant was normal.

In puzzling over this mutant, I remembered that putrescine was formed from ornithine directly and from arginine via agmatine. In the presence of arginine, ornithine would not be formed because of feedback inhibition. If the mutation is between arginine and putrescine, the bacteria would require putrescine, provided the polyamines, such as putrescine or its product spermidine, are required for growth. This turned out to be the case. Putrescine (and spermidine) overcame the inhibition by arginine.

The finding of this mutant started a new line of research in our laboratory. Together with two postdoctoral fellows, I. N. Hirshfield and H. J. Rosenfeld, and a graduate student, Z. Leifer, I started to investigate the genetic basis and function of polyamines. The mutant I had isolated was blocked between agmatine and putrescine, with a greatly reduced amount of the enzyme agmatine ureahydrolase. Subsequently, we isolated mutants in all the other steps in polyamine synthesis and mapped the locations of the corresponding genes in the *E. coli* chromosome. They were clustered together in a small region.

As with other mutants, we thought they would be useful in studying the specific functions of polyamines, but we were not successful. It may be that polyamines, being organic cations, have many functions like inorganic cations, such as magnesium (Mg^{++}). The only clear defect we could see in these mutants was the formation of long "snakes" due to a lack in septation. After about five years, I gave up working on polyamines.

Chapter 9

The Arginine Regulon

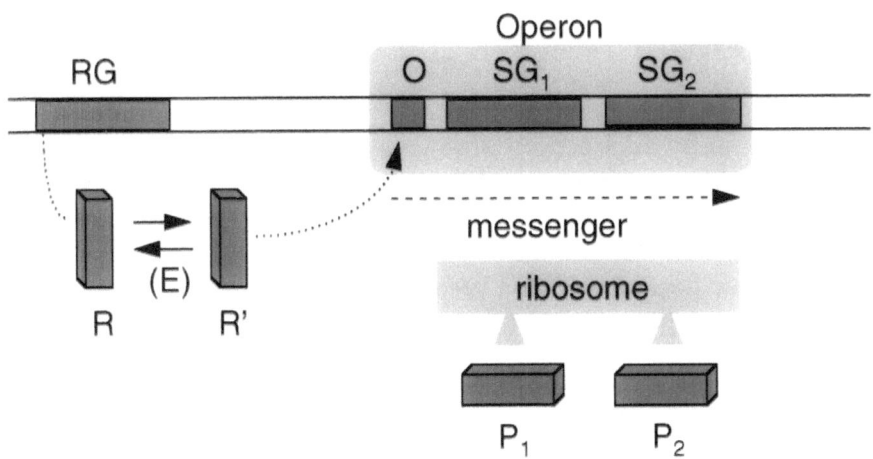

Figure 1. The Operon Model. The Repressor R is converted to R' by combining with a co-repressor E. It acts on the operator O to control the activity of the structural gene SG_1 and SG_2, responsible for the production of proteins P_1 and P_2.

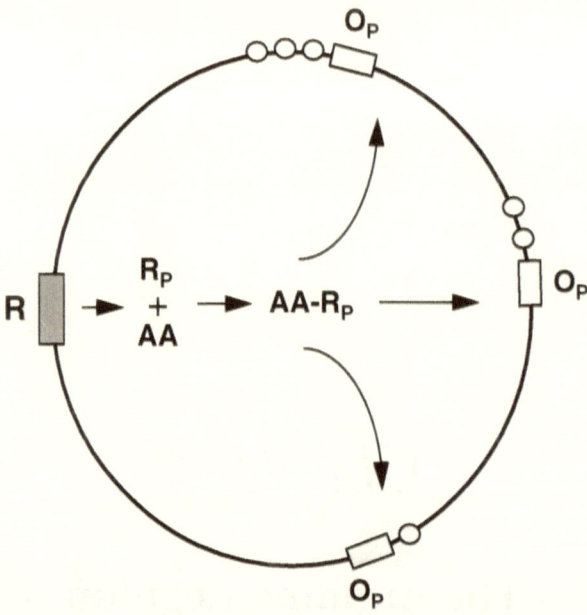

Figure 2. The Regulon Model. A repressor molecule Rp, product of the gene R combines with the amino acid AA whose synthesis is controlled by three genes. Each of these genes constitutes an operon, with an operator Op.

By 1960, the steps between genes and their protein products had been sufficiently analyzed to permit a study of the site of action of repressors. A model for the action of repressors, called the Operon Model, was presented at the 1961 Cold Spring Harbor Symposium by Jacob and Monod based on their studies of the lactose system (fig. 1). The evidence for the model is based partly on the Pajamo experiment described in chapter 6. In addition, Jacob isolated cis-acting mutants that eliminated the response to the repressor. In mapping experiments, he showed them to be located at the beginning of the operon at a site called the operator. These mutations affected the action of all the genes of the lactose operon. This finding suggested that the repressor acted directly at the operator, inhibiting the formation of the recently discovered messenger RNA for the lactose genes, as shown in figure 1. The operon is thus a genetic unit of regulation. A similar arrangement of linked genes regulated by the action of a repressor on a single operator was found in the biosynthesis of histidine and tryptophan.

Since the arginine genes are scattered rather than linked, I decided to carry out a Pajamo experiment using the regulation of OTCase as an example to test the generality of the Operon Model. At that time, we had established a collaboration with a group at the Free University of Brussels under the direction of J. M. Wiame. He and his student, Nicolas Glansdorff, came to New York to participate in the execution of this experiment.

In order to carry out the experiment, we needed to have an OTCase negative mutant, but such a mutant was not available in *E. coli* K12, the strain we had chosen for our experiment. It was, however, available in the *E. coli* B. We therefore had to transfer the mutated gene by mating *E. coli* B with *E. coli* K12. I had recently married Renata Diringer, who was trained as a biochemist and who asked me if she could carry out this transfer to gain some experience in microbial genetics. There were some technical difficulties in mating between the two strains, but she finally succeeded in the transfer by using an F-Lac$^+$ *E. coli* B as donor and an Hfr *E. coli* K12 as recipient. (Hfr stands for high-frequency donor.)

In the mating that was then carried out between a nonrepressible Hfr strain that was ArgR$^-$ OTCase$^+$ and a repressible recipient strain that was ArgR$^+$ and OTCase$^-$, the zygotes formed were repressible. As in the Pajamo experiment, repressibility was a dominant trait.

As in the case of the lactose operon, I also obtained, with the collaboration of Alvin Clark, a permanent diploid strain that carried an ArgR$^+$ gene on the chromosome and an ArgR$^-$ gene on a plasmid. In this strain, the ArgR$^+$ phenotype was dominant, thus confirming the results obtained with the temporary zygotes.

As shown in figure 1 of chapter 6, the genes of arginine biosynthesis occur in one cluster of four genes and four single genes. I assumed that there are five operons for arginine biosynthesis and they are all controlled by the arginine repressor. I coined the term *regulon* to describe a physiological unit of regulation that may include more than one operon, all under the control of the same repressor (fig. 2).

At the time of these experiments, no cis-dominant operator mutants analogous to such mutants in the lactose operon had been isolated. Several years later, such mutants were found for the Arg ECBH cluster and the ArgF and ArgI genes. The final proof for five separate operons came from sequencing studies (see chapter 10).

As a final comment in this chapter, one may wonder why no OTCase mutants could be isolated directly in the *E. coli* K12 strain. It turned out that there are two genes for OTCase production in this strain, ArgF and ArgI, an anomalous situation. In our transfer experiment of the OTCase gene from *E. coli* B to *E. coli* K12, the incoming chromosome from *E. coli* B eliminated both OTCase genes of *E. coli* K12.

Chapter 10

The Molecular Mechanism of Repression in Arginine Biosynthesis

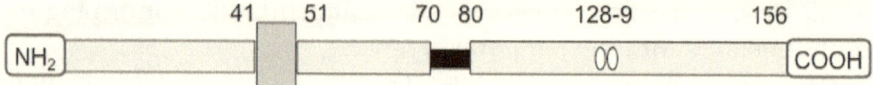

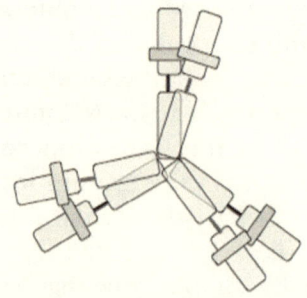

Figure 1. Schematic presentation of the structure of the arginine repressor molecule. Above: A monomer, showing the binding site for DNA in the N-terminal domain at position 41-51 and the arginine binding sites in the C-terminal domain indicated by circles. Below: The two superimposed trimers forming the hexameric repressor molecule.

Once the arginine regulon was established in physiological and genetic studies, further work was directed toward investigating the chemical nature of the arginine repressor (both aporepressor and corepressor) and the operator sites.

In chapter 7, I mentioned our studies in 1969, dealing with the possibility that the corepressor was arginyl-tRNA rather than arginine. This possibility was ruled out by several experimental approaches. We showed that in mutants defective in arginyl-tRNA synthesis, repression of arginine synthesis was normal. Moreover, in these mutants as well as in the wild type, the level of charging of the 5-arginyl-tRNA species with arginine was the same under conditions of repression and of derepression. Finally, in experiments on the cell-free synthesis of the enzyme arginosuccinase, carried out in collaboration with H. L. Yang and G. Zubay of Columbia University, arginine repressed enzyme synthesis in an extract that was freed of arginyl-tRNA synthesis. In these same experiments, we showed that the extracts had to be prepared from an ArgR$^+$ strain. In an extract of an ArgR$^-$ strain, there was no repression by arginine. My associate Norman Kelker purified the repressing substance about seventy-fold by conventional methods of protein preparation. At this point, we assumed that the aporepressor was a protein. This assumption was strengthened by the results obtained in collaboration with the group in Brussels, showing that in purified repressor preparations, arginine inhibited the formation of ArgCBH messenger RNA. These results also showed that the repressor acts at the level of transcription.

A great advance in our studies occurred after methods for cloning and sequencing DNA were developed during the 1970s. After the whole *E. coli* K12 genome was sequenced, it was shown, mainly by the Brussels group, that at the beginning of the genes for arginine biosynthesis and the repressor gene ArgR, there were two "Arg boxes" each consisting of an eighteen-base pair palindromic sequence. They were the operators containing the binding site for the repressor. The Arg boxes in the different genes were similar but not identical. In some cases, they overlapped the start site of transcription (promoter).

In the 1980s, a Swiss postdoctoral fellow, Thomas Eckhardt, cloned the ArgR gene. By using restriction enzyme analysis, he localized the gene on a two-kilobase fragment. He and a newly arrived graduate student from Korea, Dongbin Lim, sequenced the ArgR gene. Dongbin then demonstrated that the ArgR gene is autoregulated. This was done by constructing an arg-lac fusion gene and showing that β-galactosidase formation is inhibited in the presence of an ArgR$^+$ allele and arginine. He also showed in primer extension experiments that the control region of ArgR contains two promoters, one of which overlaps with two arg boxes and is inhibited by the ArgR protein and arginine.

In studies on the repressor protein, Dongbin first isolated the Arg-lac fusion protein and obtained antibodies against it. These antibodies were useful for monitoring the repressor during its purification. By putting the ArgR gene under the control of the potent tac promoter and deleting the normal promoters, he was able to hyperproduce the ArgR protein and to isolate the repressor in pure form. This was greatly facilitated by the finding that the repressor protein could be selectively precipitated from crude extracts by the addition of arginine. He showed that the native repressor protein is a hexamer with a molecular weight of ninety-eight thousand. Each of the identical subunits contains 156 amino acids. He showed in in vitro runoff transcription experiments with ArgF, the gene for OTCase, that the repressor inhibits transcription in the presence of arginine.

Following the purification of the repressor, the interactions of the repressor with operator sites of the arginine regulon were investigated in detail by Dongbin and Guoling Tian, a Chinese graduate student. The methods used were various kinds of in vitro footprinting and gel retardation. These experiments showed that the repressor binds simultaneously and cooperatively to two adjacent Arg boxes on one side of the double helix. The affinities for Arg boxes, as measured by half-maximum protection, range from 10^{-9} to 10^{-10} Molar, with ArgF having the strongest affinity.

Gel retardation experiments with operator fragments of several Arg genes gave results for binding activities similar to those obtained in the footprinting experiments.

In the 1990s, we began to study the molecular structure of the repressor. To begin with, Guoling isolated and mapped mutations of the repressor to localize the binding sites for DNA and for arginine. The mutants she characterized were of two kinds. One, based on resistance to canavanine was defective in repressor production. The twenty-two mutants she isolated mapped throughout the repressor gene. Mutants located in the N-terminal part of the molecule were defective in the binding to DNA. Mutants located in the C-terminal part were defective in the binding to arginine. Some of these could not form hexamers. This gave us a rough idea about the domain structure of the repressor.

The other kind of mutants, selected for requirement of arginine for growth, were superrepressor mutants producing more potent repressor molecules than the wild type. She mapped twenty-seven of these, and they also occurred throughout the gene. Some of them had greater affinity for DNA in the absence of arginine, others only in its presence.

They did not tell us anything about the structure of the repressor, but they were of great interest for studying its mode of action.

In the 1990s, we began collaborating with Paul Sigler at the University of Chicago to study the structure of the arginine repressor by x-ray crystallography. By that time, it had become possible to clone the two domains of the repressor separately and to isolate the corresponding proteins. Paul assigned the project to his talented student Greg Van Duyne who managed to isolate crystals of the C-terminal arginine-binding domain that were suitable for obtaining a high-resolution picture of this part of the repressor. It showed when and how arginine bound to the repressor and the arrangement of the six subunits into two trimers sitting on top of each other.

We isolated and studied the N-terminal protein in collaboration with Jannette Carey at Princeton. We showed that a monomer could bind to operator DNA and could repress OTCase formation in the absence of arginine. Jannette subsequently collaborated with some Swedish and German structural biochemists to determine the structure of the N-terminal domain by NMR spectroscopy. It was found to be of the hinged helix-turn-helix type commonly found among regulatory proteins. There were three helices, one of which bound to DNA.

It now became possible to make a model of the overall atomic structure of the repressor with the N-terminal domain with seventy amino acid residues and the C-terminal domain seventy-six amino acid residues, the two joined by a hinge region of ten residues (fig. 1).

Early in the 1990s, Paul Sigler and his group had moved to Yale University where the work on the arginine repressor was continued. In 1998, after obtaining a PhD degree, Greg became a faculty member at the University of Pennsylvania where he maintained his collaboration with our laboratory. At that time, he had the idea that changing the arginine binding site of aspartic acid at position 128 to asparagine would lead to the binding of citrulline to the repressor rather than arginine. We isolated a mutant with asparagine at position 128, and his prediction turned out to be correct. There was, however, an unexpected effect. The mutant with asparagine turned out to be a strong superrepressor by itself. Mutants with the substitution of other amino acids for aspartic acid had no such effect. Neither had the substitution of amino acids (including asparagine) for aspartic acid at position 129, which is another arginine binding site.

Since superrepressors are characterized by requiring arginine for growth, we looked for mutants that had lost the arginine requirement and presumably were no longer superrepressors. We found such a mutant, and it mapped in the hinge at position 76, changing serine to proline. This mutant, transferred to the wild-type strain, had no effect on the requirement of added arginine for repression.

What these results illustrate is the flexibility and versatility of the repressor molecule in determining its binding to DNA. As we have seen, the same kind of mutants can be brought about by single amino-acid changes throughout the repressor molecule and depend on the nature of the replacing amino acid. So far, we know very little about changes in operator DNA as a result of binding of the repressor and how such changes interfere with transcription by DNA polymerase. Greg has now crystallized repressor protein linked to Arg box DNA and is studying the structure of the complex. This is the beginning of investigating the question I posed about the mechanism of repression that remain to be answered.

Finally, I should mention that during the last twenty years, arginine repressors with similar hexameric structures have been found and described in diverse bacterial species, including *S. typhimurium, B. subtilis, B. stearothermophilus, H. influenzae,* and *M. tuberculosis.* This widespread occurrence suggests that the arginine repressor is of ancient origin.

Chapter 11

E. coli B, its Evolution
and its Natural Selection

During the 1980s, Dongbin carried out the same procedures with the arginine repressor of *E. coli* B as he had done previously with the *E. coli* K12 repressor. He cloned and sequenced the ArgR gene, and he isolated the repressor protein in pure form.

There were five differences in the DNA sequence between the two repressor genes, but only one amino acid difference between the repressor proteins, proline at position 70 in K12 was replaced by leucine in B. This difference resulted in the B repressor being much less soluble in the buffer used during purification than the K12 repressor, but both could be precipitated by arginine. In addition, ArgR⁻ mutants of *E. coli* K12 and *E. coli* B had the same high derepressed level of OTCase.

It took several years before Dongbin and Guoling could work out how the single amino acid difference between the repressors could bring about the altered response to the addition of arginine during growth. They found that the B repressor binds quite well to Arg box DNA in the absence of arginine, whereas the K12 repressor requires arginine for effective binding. This explained why the synthesis of the arginine biosynthetic enzymes in *E. coli* B is maintained a constant low level. In fact, the B repressor can be considered to be a weak superrepressor. Among the superrepressor mutants isolated by Guoling, there was a strong one at position 70 in which proline is replaced by serine.

During recent years, I learned where and how the arginine repressor of *E. coli* B evolved. Before describing the experimental evidence, I should mention that *E. coli* B was isolated from sewage by Jacques Bronfenbrenner at Washington University in St. Louis and brought to the Cold Spring Harbor Laboratory by his associate Alfred Hershey, where it was used by Luria and Delbrück in their "classical" experiment and after that became a standard strain for research.

The question of which came first in evolution, ArgR^K or ArgR^B, was investigated by Antony Dean and his associates at the University of Minnesota. The ArgR genes of eighty-five strains from various sources were sequenced, and nineteen polymorphic sites were found of which seventeen were silent (no amino acid changes), and two of them represented amino acid replacements. One of these, the ArgR^B mutation, was unique for the *E. coli* B strain. The other, a change from aspartate to glutamate at position 143, did not affect the ArgR^K repressor function.

There were sixteen unique haplotypes among the ArgR sequences. The ArgR^B sequence was present only in one haplotype. In the derived haplotype network, this haplotype is in an isolated position and differs from its connecting haplotype only by the proline to leucine change at position 70. This fact together with its low frequency of occurrence indicates that ArgR^B is of recent origin and evolved from a strain with ArgR^K.

The Dean group has also investigated the question of what kind of environment favors an *E. coli* strain with the ArgR^B repressor. They carried out chemostat experiments in which they determined the outcome of competition between *E. coli* K12 and *E. coli* B under various conditions. They found that the presence of arginine favored the growth of *E. coli* K12. The absence of arginine favored the growth of *E. coli* B. This finding is in accord with our previous finding that in a heterozygous diploid strain, B-type regulation is dominant over K12 type whereas in the presence of arginine, the reverse is true.

In a general way, these findings demonstrate that the strategy of regulation that is selected depends on the nature of the environment.

I was pleased by the results obtained by the Dean group. They showed me that our research, besides describing a basic mechanism of gene regulation, provided the means for analyzing the role of regulation in natural selection. It was this kind of understanding of the relationship between genetics and evolution that lead Theodosius Dobzhansky, the mentor of my PhD project at Columbia University, to make his profound statement that "nothing in biology makes sense except in the light of evolution."

Chapter 12

Looking Back

It was a lucky coincidence that the 1941 timing of my decision to study the biochemistry of gene expression, after having taken courses in genetics and biochemistry, coincided with the publication of the groundbreaking paper by Beadle and Tatum that launched the field of biochemical genetics, which later became molecular biology. I came in on the ground floor. I realized, at the time, that I had to have a good background in both biology, especially genetics, and in chemistry. I tried to acquire that during five years of graduate and postdoctoral studies at Columbia and Caltech.

When I started to do serious research in Bernie Davis's laboratory, I found out that doing biological studies translated into in vivo experiments and chemical studies into in vitro experiments. For in vivo experiments, mutations turned out to be very useful tools, and I tried to obtain appropriate mutants in all my projects. I referred to them as living dissection needles.

Looking back, I arrived at answers to the questions I had posed to myself about gene expression and its regulation through my curiosity about apparently unrelated findings. For example, my finding that D-serine was inhibitory, when in 1948 I tried to grow the serine-requiring, temperature-sensitive mutant, made me pursue the mechanism of inhibition by D-serine rather than look for the enzyme affected by the mutation. After I had determined that D-serine inhibits the last step in the formation of pantothenic acid, I tried to characterize this inhibition by studying it with the extracted enzyme. I also obtained a pantothenate-requiring mutant for the assay of pantothenic acid. I was successful in these efforts, and it was then that I accidentally found the derived temperature-sensitive mutant that made it possible to demonstrate that the mutant produced an altered enzyme.

In connection with the approach, I took to find a solution for my problem, I would like to quote a remark Fritz Lipmann made when asked by reporters about the secret of his success leading to a Nobel Prize. He smiled and said, "Oh, I just follow my nose." I found this to be an apt description of my own pursuit of my problem.

In 1954, after the elucidation of the structure of DNA, work on the biochemistry of protein synthesis from DNA was in full swing. At the time, I did not join these efforts but rather started to investigate the, to me, more interesting question of the regulation of gene expression. Here again, I used a mutant approach, this time in arginine biosynthesis, expecting that a mutation resulting in cold sensitivity of an enzyme was due to a change in gene regulation.

It was a "just in case" control, the addition of arginine at thirty-seven degrees, that showed that the end product of the pathway was responsible for the regulation of the enzyme OTCase, which I was using. As the same inhibition was found in the wild-type strain, the use of the cold-sensitive mutation was irrelevant. I had discovered a general process.

My further studies on the regulation of enzyme synthesis in the pathway of arginine formation opened for me the new vista that there is no structural relationship between the regulating substance and the enzyme whose production is being regulated by this substance. Instead, the response to arginine, the regulating substance, is mediated by the product of a gene whose sole function is the regulation of gene expression, in my case, a repressor for the nine genes of arginine biosynthesis. In a general way, I had discovered that besides genes that determine the structure of cellular constituents, there are regulatory genes whose product controls the rate of synthesis of arginine in response to environmental conditions. As I stated in a review in 1974, the cell was no longer considered solely as a chemical machine but as a cybernetic chemical machine.

After 1960, my efforts for the next thirty-five years were directed toward working out the biochemical mechanism of arginine expression. At the end, I arrived at a fairly concrete picture of the process, at least for the repressor molecule and how it binds to DNA. What was unusual was the hexameric structure of the repressor protein, consisting of two sets of opposing trimers. What I found most impressive was how mutational changes in different parts of the repressor gene could affect the binding to DNA, thus altering the action of the repressor. As I demonstrated in chapter 11, the importance of regulatory genes in evolution, the great variety of allosteric effects on repressor binding due to mutational changes provides many possibilities for response to natural selection.

Finally, looking back at my efforts to investigate experimentally the questions I had posed to myself after my courses in genetics and biochemistry, I find that what was most helpful was to keep an open mind in the interpretation of experimental findings. This made me carry out control experiments that at times seemed excessive, but they prevented me from straying off in the wrong direction—away from the path that nature had taken in the course of evolution.

www.ingramcontent.com/pod-product-compliance
Lightning Source LLC
Chambersburg PA
CBHW021926170526
45157CB00005B/2200